SAFE OPERATING PRACTICES

BY BIPLAB ROY CHOUDHURI

1. CANTILEVER SKIDDING

1. Rig crew should use proper PPE.
2. Rig crew should check the disconnection of hoses.
3. Rig crew should secure the loose item.
4. Rig crew should give proper signal.
5. Rig crew should check the drill pipe strand before skidding.
6. Rig floor holes to be covered.
7. Rig crew should check for any obstruction in platform with Rig floor or any other part.
8. Rig crew should apply grease and avoid slipping while applying grease.
9. Rig crew should watch for any obstruction in the way to movement.

2. RIG FLOOR SKIDDING

1. Rig crew should use proper PPE
2. "Same as above"

3. BOP & ANNUAR TEST

1. Rig crew should use proper PPE.
2. Rig crew should give proper signal for high Pressure Test.
3. Rig crew should watch for any leakage.
4. Rig crew should stay away from High Pressure Line.
5. Rig crew should check for annulus valve opening secure the loose item.
6. Testing authority to signal when testing is complete.
7. Rotary Table should be clear of all personnel.
8. During testing Rig floor will be clear of all personnel.
9. Cellar area should be clean and clear of obstacles
10. When H^2S is present all personnel should have H^2S Certificate.
11. Testing area should be covered by a barricade tape.
12. Casing Valve should be open and not plugged.
13. Fluid should not leak from casing valve.
14. D/P annulus should be checked for leakage from test plug.

4. General Safety on drill floor

1. Crew should put on Helmet, Safety shoe, Goggles, Gloves.
2. Floor area should be clean.
3. Floor area should not be slippery.
4. Winch hook should be secured.
5. V door should be closed.
6. Driller should be on brake.
7. If driller not on brake it should be secured.
8. Floor space should not be congested.
9. Winch line to be checked periodically.
10. Crowno -matic to be checked on regular basis.
11. Crew should not keep anybody part near survey line while operating winch.
12. Survey tool crew should not try to hold it while in operation
13. oil is the Lubricant spillage need to be cleaned immediately.

14. While stabbing pipe/Tool crew should not place hand in improper position.
15. Crew should not be near to heavy load while lifting.
16. While using snatch block crew should be care full.
17. During operation of Winch crew should give signal and operator should start operation only after getting signal.
18. While holding the rope to secure it should be tied properly at both end.
19. While lifting heavy load, the rope tied with the material should be hold tightly.
20. When testing high pressure cementing line alert should be given and crew should stay away of rotary table or testing lines.
21. Proper communication should be tween top man, driller and derrick man.
22. Mast should be checked for loose stud,.
23. In case of burnt of any oil or lubricant line, crew is advised to keep away immediately from that place.
24. Proper clothing should be used by rig crew no loose clothes are allowed.
25. Hanging pendant should not be allowed.
26. Derrick ladder condition to be checked periodically.
27. Corrosion of derrick to be checked & replaced thereafter.
28. Fall protection device to be checked.
29. Pipe finger should be in good condition.
30. Walk platform in derrick board should be in good condition to avoid slip and fall.
31. All Rotary chain drives to be guarded.
32. All floor hole to be covered.
33. Air Winch line should be properly wrapped.
34. Rig floor area should be in good condition.
35. Slip and die should be in good condition.
36. Wire line should not be worked out or kinked.
37. Wire line Unit to be secured.

38. Rope and snatch block should be of correct size & strength(SWL) marking.
39. Rope inspection before job.
40. Slices should not be used in the entire length of rope.
41. High pressure line should be checked for leakage periodically.

5. **Nipple up down of BOP**

    1. Crew should put Helmet, Safety glass, Gloves, Steel Toe Shoe.
    2. Pick up Winch line should be checked.
    3. If required hot work permit is to be taken.
    4. Keep fingers, hand feet and other body part away from bolt and flange.
    5. All the bolts are used with full nut.
    6. Secure hammer, wrenches, of Torque wrench not used.
    7. Be aware of body position while tightening or breaking of studs.
    8. All hammer wrenches should be clean.
    9. Do not place Body part under hanging BOP.
    10. For stabbing use Air Winch.
    11. Do not use sharp edge with slings while positioning.

6. **Nipple up and down diverter**

    1. Crew should use Helmet, Safety Glass, Gloves, and Shoe.
    2. Check for Winch line and Winch.
    3. Finger, Hand feet and body part away from bolt.
    4. Failure to open/close valve as required when subject to gas flow.
    5. Erosion of internal surface to be checked.
    6. Valve should be capable of opening with max anticipated pressure.
    7. Diverter pipe erosion and pressure drop are major concern and to be checked.

8. Flexible piping should not be used Diverter vent line should be hard piping.
9. Diverter pipe should be straight.
10. Vent line to be secured.

7. **Nipple up Well Read Section A**

1. Crew should use required PPE.
2. Hot work permit is required prior to cutting casing.
3. Welded area not to be touched by naked finger.
4. Proper signalling should be given from well head fitting team to driller.
5. Proper welding should be done.
6. Well head to be tested.

8. **Nipple up Well Head B & C Section**

1. Crew should use PPE
2. Fingeror hand should not be in bottom of flange.
3. Proper signalling should be established by Well Head team to Driller.
4. Positioning of stud should be done with two studs.
5. Hammering technique in restricted area should be discussed.
6. Deck area should be cleaned.
7. Sufficient tightening of stud should be there.
8. Air Winch should be used while fixing, if necessary.
9. Hot work permit should be there.

9. **Nipple up DSA on Well Head**

1. Rig crew should put on proper PPE
2. Crew should not place finger or hand below flange.
3. Crew should give proper single to Winch operator.
4. Deck area should be clean and it should not be slippery.
5. Rig crew should orient DSA properly.

10. **Choke & Kill Manifold Test**

    1. Crew should use the proper PPE.
    2. Floor area should not be slippery.
    3. Testing should be done on all the valves in proper sequence.
    4. Crew should be away from Testing Lines.
    5. Testing Supervisor should announce the testing.
    6. Testing Supervisor should announce the completion of Testing.
    7. Choke and kill manifold pressure Test not exceed 50% of the working pressure.
    8. Line should be checked for leakage, flange and seals are subject to checking.
    9. If leak, release pressure and rectify leaks.

11. **Stand Pipe Manifold Test**

    1. Crew should use proper PPE
    2. Rig floor area should be clean.
    3. Testing should be done on all the valves in sequence.
    4. Crew should be away from testing area.
    5. Testing supervise should announce Testing.
    6. Testing supervisor should announce completion of Testing.
    7. Testing line should be checked for any leakage.

12. **Cmtg Manifold Test**

    1. Cmtg. Engr. Should use proper PPE.
    2. Cmtg Unit should be clean and not slippery.
    3. Cmtg. Unit should be cleaned.
    4. Cmtg. Operator should announce for Testing.
    5. Cmtg. Operator should announce for completion of Testing.

6. Rig crew should stay away from testing zone.
7. Rig crew should check for leakage from cmtg line.
8. Cmtg. Operation should watch for pressure drop.

13. **BOP Test with Test Unit**

    1. Rig crew should use proper PPE.
    2. Tool should be properly cleaned.
    3. Rig crew should be careful as job will be done near to rig.
    4. Rig crew should check for different work condition.
        - Night Time Operation
        - Day time operation
        - Hot climate
        - Cold Climate
        - Wet weather condition.
        - High wind.
    5. Rig crew should check for procedure for nippling up of BOP and pressure testing.
    6. First hand testimonial evidence of crew working with other to undertake and complete nipple up and pressure testing.
    7. Rig crew should give proper instruction and signalling.
    8. Proper communication should be there.

14. **BOP Ram Change and Packer**

    1. Ensure that no one in on or near BOP when opening or closing the BOP bonnet.
    2. BOP and work area should be clean and clear of obstruction.
    3. Blind Ram should be closed at all times.
    4. Ensure well bore is full of mud.
    5. Ensure adequate lighting at night.
    6. Ensure good communication between crew.
    7. Hold Hammer Wrench with rope keep hand and finger clear.

8. Inspect all lifting equipment to ensure it is in good working order.
9. Ensure Ram eyebolt fits properly and is fully screwed in.
10. Crew should instruct air hoist operator to pull up ram slowly, while someone guides thro ram body off the ram shaft using the tail rope.
11. Crew should clean the ram cavity and check for wear or tear.

15. Pick up Tubular from v door

1. Crew should use proper PPE
2. Manila Rope should be tied up onV door of Rig floor.
3. Lifting protector should be fitted on pipe and should be tightened.
4. Pipe should be picked up by winch smoothly.
5. Rope should be guiding the pipe and slowly bring to Rotary.
6. Winch Line should be properly wrapped not loose.
7. Thread protector should be opened.
8. Manila Rope to be slacked off properly.
9. It Latching by elevator care should be taken so that hand finger,
   should be clear away.

16. **Pick up BHA**

1. Crew should use proper PPE
2. Drill collar should be lifted very slowly and latched with elevator.
3. Rig floor holes to be covered.
4. Crowno-matic should be set for Drill Collar strand pulling out.
5. Winch to be used for guiding Drill Collar.

6. While lowering drill collar in hole proper size safety clamps to be used and properly tightened.
7. BOPS shear Ram to be closed.
8. Drill Collar to be tightened by chain tong and finally by manual tong.
9. Non-magnetic drill collar with bent sub to be tightened very carefully without cross threading.
10. Rotary Lock not to be used.
11. Proper type of thread compound to be used.
12. For new BHA Drill Collar Thread to be cleaned with diesel and Bentonite and dope will be applied.
13. For stabilizer Tong law should not be on stabilizer blade.
14. BHA strand should be properly racked back on wooden platform

17. **Lay down Tubular from V door**

1. Rig crew should use proper PPE.
2. Rig crew should properly tight lifting cap.
3. Rig crew should check winch line.
4. Air Winch operation should be smooth.
5. No person should be on catwalk.
6. Rig crew should put Thread protector on pipe.
7. Lay down pipe should be placed on catwalk.
8. Finger/Feet should be away from drill pipe.
9. Pipe should be rolled using wood.
10. Shackles should be checked for hand tight.

18. **Lay down BHA**

1. Rig crew should use proper PPE.
2. Rotary hole should be covered after breaking bit.
3. Tong to be placed properly.
4. Tong dies, slip dies should be proper.
5. Air tagger line should be used properly for guiding BHA.
6. While laying down Drill Collar care should be taken.
7. Rig floor should be adequately cleaned.
8. Floor should not be slippery.

9. Rotary Holes to be covered.
10. Crew should not attempt to stop or step over a rolling tubular.
11. Crew should stand clear when driller lower collar into trough and elevator are unlatched from lift sub.
12. Rig crew should put rope on V door stair, catwalk area and other hazards.
13. There should be proper lighting catwalk V door.
14. While breaking out drill collar crew should stay clear.
15. Pipe rack should be fitted with end stopper.
16. Rig crew should not walk on d/p or drill collar.
17. Crew should stand clear on ground while point is laid down by lay down operator.
18. Crew should inspect condition of hoist and cable they should know the capacity of cable and hoist.

19. **Handling tubular**
    1. Rig crew should put on proper PPE.
    2. Rig crew should inspect sheaves cable.
    3. Rig crew should tie rope in v door stair, catwalk area.
    4. There should be proper lighting on pipe rack, catwalk and V door.
    5. Rig crew should not walk on casing, tubulars, they should use the stairway.
    6. Rig crew should move only one joint of casing.
    7. Crew should not use hand or feet to move pipe on top of pipe rack.
    8. Crew should stand clear when lifting pipe to rig floor.
    9. Crew should use the casing protector.
    10. Crew should not attempt to stop or step over rolling tubulars..

20. **Installing lifting sub on DC on V door or deck.**
    1. Rig crew should use proper PPE.
    2. Lifting sub should be filled properly on drill collar.
    3. Rig crew should check the drill collar while picking up.

4. Rig crew should check the lifting sub condition while lightning the bit by rotary tong or chain tong
5. Lifting sub wear on thread be checked.
6. While checking Torque gauge driller should check lifting sub
7. Rig crew should not turn, collar until person is watching lifting sub.
8. Rig crew should check collar box and lifting sub thread.

21. **Installing lifting Cap on HWDP/DP or V- door or deck**

1. Rig crew should use proper PPE
2. Lifting cap should be lightened properly.
3. Crew should check the thread of lifting cap and HWDP Box end for wear.
4. While picking up HWDP crew should check lifting cap.
5. Crew should not use one hand in putting lifting cap.
6. When opening lifting cap crew should check winch line is taken off.

22. **Break out Drill Collar**

1. Rig crew should use proper PPE.
2. Rig crew should keep two or three drill collar above while breaking out bit.
3. One person should watch the lifting sub.
4. Crew should not stand in the radius of snub line of makeup Tong.
5. There is no need to touch the Tong elsewhere except handle to place the Tong on Drill collars.
6. When breaking out or applying Torque to drill collar, Rig crew must stand clear of the area.
7. Tong parts like pin, jaw should be of correct size and rating.

23. **Racking Tubular in finger board**

1. Crew should use proper PPE.
2. Tubular should be pulled back by using Manila Rope properly.
3. Communication should be between derrick man and driller.
4. In pipe rack system derrick man should close the latch of pipe rack system, while engaging.
5. For remote operated finger derrick man should check finger, opening or closing.
6. Derrick man should use safety belt.
7. Crew should check finger for damage.
8. Crew should walk finger area cautiously not to fall on finger gap
9. Sufficient lighting should be on pipe rack area.
10. Crew should secure the pipe in finger first and do the next operation.
11. Crew should be avoiding falling while climbing up or down the ladder.
12. Crew should avoid falling from monkey board or finger.
13. Crew should avoid being caught between pipe and object.
14. Crew should be alert from recurring strain and sprains.
15. Crew should avoid being stuck by dropped object.
16. Crew should not carry tool, while climbing derrick ladder.
17. Crew should be carefully not to getting hands and fingers between stands of pipe.
18. Crew should avoid getting feet or toe crushed under stand of pipe.
19. Crew should inspect and maintain elevator.

24. **Racking drill collar and HWDP in finger boards**

1. Crew should use proper PPE.
2. Tubular should be pulled back by using Manila Rope properly.
3. Communication should be between derrick man and driller.

4. In PRS system derrick man should close the latch of pipe rack system, while engaging.
5. For remote operated finger derrick man should check finger, opening or closing.
6. Derrick man should use safety belt.
7. Crew should check finger for damage.
8. Crew should walk finger area cautiously not to fall on finger gap
9. Sufficient lighting should be on pipe rack area.
10. Crew should secure the pipe in finger first and do the next operation.
11. Crew should be avoid falling while climbing up or down the ladder.
12. Crew should avoid falling from monkey board or finger.
13. Crew should avoid being caught between pipe and object.
14. Crew should be alert from recurring strain and sprains.

25. **Tripping Pipe**

1. Rig crew should use proper PPE.
2. Rig crew should check safety line ropes and air host.
3. Rig crew should check the elevator latch, identify fault and report.
4. Rig crew should pick up slip without straining.
5. Rig crew should be cautious of getting finger or other body part pinched between slips or slip handle and rotary table.
6. Crew should use proper hand placement when setting slip and use proper stance.
7. Crew should be careful while attaching elevator or being stuck by elevator and stay away through swing path of elevator and elevator link.
8. Crew should avoid getting hand or finger pinched on elevator.
9. Crew should inspect & maintain elevator.
10. While tripping crew should clean floor are to avoid slippage.

11. Crew should be careful so that sprinkled muds do not fall on eye.
12. Crew should avoid Tong Line dangerous area.
13. Crew should ensure that they know proper elevator latching.
14. Crew should check for crownomatic before tripping and flooromatic devices.

26. **Run in Hole DP**

1. Rig crew should use proper PPE.
2. Rig crew should clean the floor are and it should not be slippery.
3. Rig crew should check Tong Safety Lines.
4. Crew should check slip dies Tong dies.
5. Crew should inspect and maintain elevator.
6. Crew should be away of danger zone when making up or Breaking up of pipe.
7. Crew should use proper stance while lifting slips and not to strain unnecessarily.
8. Crew should be away from Tong when torqueing or breaking out.
9. Crew should place Tong on Tool joints.
10. Crew should not use shear ram when running in pipe.
11. Crew should have proper communication between them.
12. Crew should be alert if elevator Link sways while running in.
13. Crew should be away from Rotary table.
14. Crew feet should not be touched by slip handle.
15. Crew should place hand in elevator handle while opening elevator not to use finger on any other parts.
16. Crew should check for crownomatic before tripping and flooromatic devices

27. **Pulling out of Hole (P.O.O.H)**

1. Crew should use proper PPE.
2. Crew should check slip, Tong dies.
3. Crew should keep floor area clean.
4. Rig floor should not be slippery and should be tidy.
5. Crew should avoid staying in the danger zone of tong when making or breaking out tubular.
6. Crew should pump slug and check for back flow.
7. There should be proper communication between Rig crew.
8. In open hole pulling out should be slow.
9. Driller should not use rotary in open hole.
10. Trip tank to be monitored for loss or gain.
11. Driller should check constantly the weight indicator.
12. Crew should keep safety valve, inside BOP ready and nearby.
13. Crew should observe flow channel for self-flow.
14. Top man should give signal to driller.
15. Crew should check for crownomatic before tripping and flooromatic devices

28. **P. OOH Wiping pipe**

1. Crew should use proper PPE.
2. Crew should hold rubber against the pipe.
3. Crew should use hose water to clean the floor.
4. Crew should use Wiper all line during drill pipe pulling out as procedure so that any slip die or Tong die do not fall in well.
5. Two person should hold the rubber wiper properly.
6. Crew should check for crownomatic before tripping and flooromatic devices

29. **P.OOH Racking Pipe**.

1. Rig crew should use proper PPE.
2. Crew should check for elevator slip Tong dies.
3. Crew should be away from danger zone while making or breaking out Tong.
4. There should be proper signalling between crew.

5. While Racking pipe crew should check feet or finger should not be at the bottom of the pipe.
6. Crew should check hands or fingers should not be in Racked pipes.
7. Pipe should be taken back properly and be placed on wood.
8. Crew should check rope for racking pipe.
9. Crew should check for crownomatic before tripping and flooromatic devices

30. **P.OOH with Pump**

    1. Rig crew should use proper PPE.
    2. Rig crew should check Tong slip, Elevator.
    3. Rig crew should check flow line.
    4. Rig crew should check Trip Tank.
    5. Crew should clean floor area.

31. **RIH with Scrapper**

    1. Crew should use proper PPE.
    2. Crew should not place Tong or slip on Scrapper blade.
    3. Crew should check the scrapper spring for play before Running in.
    4. Crew should replace the collapsed spring.
    5. Crew should replace blade that is worn out.
    6. Crew should completely disassembled and thoroughly clean scrapper after use.

32. **RIH with Casing Cutter**

    1. Crew should use proper PPE.
    2. Crew should check cutter blade and test cutter on surface
    3. Crew should use proper size knife for the casing
    4. During POOH casing cutter crew should check the cutter knife condition.

33. **RIH with Casing Spear.**

    1. Crew should use proper PPE.

2. Crew should check spear before running in.
3. Crew should check so that spear does not fall in the well
4. Crew should check while pulling out spear so that it does not exceed over pull limits.
5. Crew should apply Grease in spear.
6. Crew should use drill Collar or HWDP for safe releasing of spear.

34. **Fishing with Magnet**

1. Crew should use proper PPE.
2. Crew should use proper fishing magnet for hole.
3. Crew should check magnet being stuck due to line failure.
4. Crew should inspect all slings, chains pins before lifting equipment.
5. Crew should avoid getting caught in wire line.
6. Crew should operate wire line in safe speed.
7. Crew should avoid for pinching hand and finger.
8. Crew should be away from Wire Line hazard zone.
9. Crew should inspect wire line rope socket and cable head for defects before use.

35. **Fishing with junk basket** . Same as above

36. **Fishing with Overshot**

1. Crew should use proper PPE.
2. The fish should easily pass through grapple with rotation.
3. A pack off should be used if circulation is required through the fish.
4. If hole size is larger than overshot OD an oversize guide to be used.
5. If pipe is lying in a recess or against the side of well consider a hook wall guide.
6. If fish cannot be engaged consider a 'Extension Sub or milling guide.
7. Penetration of overshot in fish should be limited in order to safe releasing.

8. Under a upward strain is maintained the fishing spring shall never be rotated to right while and overshot in engaged other than when attention to come off the fish.
9. When lowering the overshot over the fish once a pressure increase is noted circulation shall be stopped in order to prevent damaging the seal.
10. After fish is engaged do not turn to right as this will release over shot.

37. RIH with Junk Mill

1. Crew should use proper PPE.
2. Crew should not rotate junk mill inside casing.
3. In floater crew should not keep junk mill inside casing in same place to avoid casing cutting.
4. Crew should check the tungsten Carbide properly brazed or not.
5. Crew should check nozzle in junk mill for circulation.

38. Installing Safety Valve/TIW Valve

1. Crew should use proper PPE.
2. Crew should keep Safety Valve handy near rotary.
3. In emergency crew should stab safety valve in drill pipe and close the valve while tripping.
4. If case of flow, crew should install safety valve and close it connect Kelly and open safety valve record shut in pressure.
5. Crew should maintain safety valve clean pin end and grease it.

39. Installation and operation Power Slip

1. Crew should use proper PPE.
2. Crew should release slip by air pressure.

3. Crew should check for leakage in airline.
4. Crew should maintain power slip properly before use.
5. Crew should check power slip and lower the sling. Crew should use foot throttle and visibly check slip
6. Driller should be very alert when using power slip and should check weight indicator constantly.

40. **Installation & Operation Spider Elevator**

1. Crew should use proper PPE.
2. Crew should pick up spider elevator slowly.
3. Crew should keep feet clear away from spider elevator.
4. Crew should check air connection for spider elevator.
5. Crew should check spider elevator should be for vertical lifting and not for any other use.
6. Crew should properly handle install and remove dies and insert to avoid hazard.
7. Crew should use eye protection at all times installing and removing dies.
8. Crew should check all pneumatic supply is isolated before any work is carried out to elevator/spider.
9. Crew should shut off the Power Unit/close the valves.
10. Crew should not use grease or pipe dope for lubricating the insert.
11. Crew should remove insert after each job, coat the in self-slot with proper fluid.
12. Crew should check for worn and damage part
13. Crew should check for loose and missing part.
14. Crew should check for sign of cracks, wear or abrasion.
15. Crew should check state of lubrication.

41. **Reverse Circulation**

1. Crew should use proper PPE.
2. Crew should be away from rotary while pumping.
3. Crew should put chickson ,swivel after pipe were run.
4. Crew should check pumping sub connection for damage.

5. As BOP is closed crew should keep watch on pressure gauge.
6. Crew should check for chain in hose connected with pumping sub so that it does not fall. (Hose or chickson or pumping sub)

42. **Working in Derrick**

1. Crew should use proper PPE.
2. Smoking should be prohibited within 25 mts of any well.
3. Open flame are prohibited within 25mt of well bore.
4. Written safe procedure must be implemented to ensure the safety of works lighting.
5. Fire extinguisher must be provided at installation.

43. **Installing and removing choke and kill lines**

1. Crew should use proper PPE.
2. Crew should install choke and kill line above bottom Ram.
3. Crew should not install choke and kill line below bottom ram.
4. Crew should check working pressure of choke and kill line match with working pressure of BOP stack.
5. Crew should check working pressure of all pipe line, Gate valve and check valve should be compatible with BOP stack used.
6. Crew should check kill manifold is not intended to be used as common lines for pouring drilling fluid.
7. Crew should check at least one adjustable choke and one manual choke should be there.
8. Crew should check chickson type choke lines are not allowed.

44. **Use care, maintenance of manual tong**

1. Crew should use proper PPE.
2. Crew should use Tong handle while biting pipe not other parts on Tong.
3. Crew should check die of manual Tong so that it does not slip.
4. Crew should check Tong /safety line.
5. Counter weight should be ok so that Tong should be in proper position.
6. While making out or breaking out crew should be away from Tong.
7. Crew should open jaws and service Tong parts.
8. Crew should use diesel while cleaning pin or jaws.
9. Crew should check jaws for proper size of pipes and correct capacity.
10. Tong die on irregular surfaces should be avoided.
11. Torque of Tong should not exceed manufacturer limit.
12. Tong parts should not be welded at site.
13. Check for missing part, worn finger pin, Grease hinge pin through grease nipples.
14. Crew should be out of Tong Travel area.
15. Tong Arm length and Tong pull Line should be at 90° Angle.
16. When applying pull, to Tongue it should be steady pull not jerking the line.
17. MPI of Tong should be carried out.

45. **Use, care, maintenance of slips**

1. Crew should use proper PPE.
2. Crew should check for corrosion, loose or missing component, detoriation ,visible external crack.
3. Crew should apply grease back of slip.
4. Crew should not put grease on dies.
5. Crew should change the die when it is worn out.
6. Crew should be careful when using slip on rotary its handle should not entangle / with feet or leg or body part.

7. Crew should lubricate the groove with light machine oil before placing the insert.
8. Crew should not weld the slip.
9. Crew should avoid getting hit by rotating slip.
10. Crew should be careful of not squeezing hand or finger between slips during maintenance.
11. Crew should check that they have placed correct insert size for the job.
12. Crew should constantly grease the slip to prevent from becoming stuck on pipe.

46. **Casing Running**

1) Crew should use proper PPE.
2) Crew should be aware of slipping and falling hazards.
3) Crew should keep work area alarm and clear of oil, Tool and debris
4) Crew should keep guard rail and guard installed around work area.
5) Crew should be cautious of not getting stuck by or caught between tubular and other object during movement.
6) Crew should be caution of moving the heavy casing tool which can result in Strains and sprains
7) Crew should be cautious of not falling from work platform or stabbing board.
8) Crew should be careful not to getting finger or body part in slip handle, elevator or stands of pipe
9) Crew should be cautious for getting stuck between Tong, casing.
10) Crew should implement a full fall protection programme for the casing stabber.
11) Crew should be cautious of not being stuck by high pressure hose during circulation or cmtg.
12) Crew should be aware of having a high pressure connection failure caused by mismatched or worn out hammer union.

47. Casing filling

1. Rig crew should use proper PPE.
2. Crew should fill up casing after interval of certain jts are run
3. Crew should fill up 1st and 2nd joint positively for checking float.
4. Crew should avoid mud spilling after casing filling.

48. Lay down Casing

1. Rig crew should use proper PPE.

2. Crew should lay down casing very slowly.

3. Crew should be aware that body part, hand feet not to be entrapped by casing.

4. There should be ample space for crew movement.

5. Crew should use more than two person for laying down casing through v-door.

6. Crew should check that casing does not swing back.

7. Crew should close all floor plate opening.

8. Crew should clear all unnecessary equipment removed from work area.

9. Crew should ensure adequate personnel/ are present to complete the task.

10. Crew should use rig laydown equipment to pull casing to the v-door slide.

49. Casing Testing

1. Crew should use proper PPE

2. Crew should Test Casing below rated pressure.

3. Alert system should be given for test.

4. Crew should check for leakage.

5. Crew should test Casing below the working rating of BOP.

6. Crew should Test Casing below well head design pressure.

7. Crew should Test casing below cement head working pressure.

8. Crew should Test casing below float equipment manufacture's test pressure

Limitation.

9. If casing is not pressure tested immediately after Bumping the plug, the casing

shall not be pressure tested again until either at least 24 Hr. of slurrry setting or

Surface sample set.

50. **Leak of Test**

1. Crew should check the surface limit pressure .
2. Crew should check for accurate pressure Gauge.
3. Crew should check pressure Rating should not exceed the burst pressure of casing and the associate surface equipment
4. Crew should not change the mud wt until after the next.
5. Crew should pump mud slowly until the pressure build up.
6. Crew should check for sharp pressure drop in surface. The highest pressure recorded, before the pressure drop is the surface break down pressure

7. If formation break down occurs pumping should be stopped.

8. Crew should make sure the hole is filled up and close the BOP around. The drill pipe where practicable open and top up the annulus between the last and previous casing.

9. Line up calibrated pressure gauges preferably mounted on special manifold. The standard pressure gauges on driller console or Cementing unit are not accurate enough for these measurements.

51. **Well shut in procedure**

1. Crew should use proper PPE.

2. Crew should notify the driller of any observed kick related warning sign.

3, A crew should assist in installing the full opening safety valve it a trip is being made.

4. Crew should include well control responsibility.
5. Crew should begin mud mixing operation.
6. Driller should shut in well it any primary kick is observed.
7. Tool pusher should notify company man properly.

52. **Well Control procedure**

1. Crew should have proper knowledge and skill

2. Crew should have knowledge of proper practices.

3. Crew should be properly trained in Well Control Training.

4. Crew should use application of policies, procedure and standard.

5. Crew should be careful of proper risk management.

6. Crew should build well Control culture which would involve developing competent personnel that are able to recognize, Well Control problem and mitigate them.

7. Crew should do programmes like Well Control accredition programme (Well cap) or IWCF Certification.

53. Using the safety clamp.

1. Crew should use proper PPE

2. Crew should check for Uniform gripping pressure so that it does not crush thin wall pipe.

3. Crew should tighten safety clamp properly.

4. Crew should check, for loose item like pin, insert.

5. Crew should place hand in proper place.

54. Slip & Cut

1. Crew should use proper PPE.

2. Crew should inspect Equipment antifall equipment, block hanging Line, shackles.

3. Crew should keep drill string at bottom of last casing string and well secured.

4. When derrick man is lifted, use tag line to position derrick man properly.

5. When derrick man is lowered use tagline to position derrick man properly.

6. Crew should avoid pinch points and store guards in safe position.

7. Crew should ensure drilling line is secured prior to cutting, to avoid and recoil in line

8. Crew should watch for pinch point. Crew should not let drilling line slide through hands.

9. Crew should properly remove the drilling line clamp and inspect for wear.

55. Rig up, Use, Care and Maintenance of casing power Tong.

1. Crew should use proper PPE

2. Crew should use winch to lift power Tong slowly.

3. There should be enough space available for using power Tong and should be enough space available for using power tong and should not be caught between power Tong and other tubular or equipment.

4. Safety line should be checked for power Tong.

5. Crew should not use power Tong with jerk.

6. Crew should use power Tong from a deck foundation and should check at for use against fall from height.

7. Crew should check for jaws for gripping.

8. Crew should check for worn out Dies, brake band or jaw size correct selection.

56. Cmtg. Operation

1. Crew should use proper PPE.

2. Cmtg operator should announce for high pressure testing.

3. Crew should fix chickson lines properly.

4. Crew should check for any leak during testing of cmtg line.

5. Cmtg Operator should flush the cmtg unit before and after the job.

6. Cmtg Operator should maintain the unit properly.

57. Working at height

1. Crew should use proper PPE.

2. Crew should use Safety belt when working at height.

3. Crew should be careful while working on height.

4. Grating, platform should be covered.

5. Handrail of platform should be installed.

6. Person with acrophobia should not be allowed to work on height.

7. There should be proper communication verbal or nonverbal between crew and supervisor.

8. Crew working involved in scaffolding and rigging work should hold certificate of competency.

9. Safety harness should be used by a person who has been trained and can perform a rescue operation.

10. Crew should be trained for reporting system related to hazards, near miss and incident.

11. Crew should be trained for prevention of fall through safe system of work.

12. Crew should check the safety harness in accordance with and be used in compliance with safety procedure.

13. Crew should take the permit to work at height duly signed by work place manager.

14. Crew should barricade below area for risk of falling objects and posting warning signs.

15. Working at height should be considered at height of 2 M or above.

58. Use, Care, and Maintenance of Utility Winch

1. Crew should use PPE

2. Crew should be trained and have adequate understanding of hoisting equipment.

3. Crew should use cable eye for easier recognition by operators.

4. Before hoisting a Winch line crew should verify the live end of the cable intended to be hoisted

5. Crew should not tie Winch cable end to equipment when not in use.

6. Crew should check cable is tied on, pull it on the winch see if the end moves then they have operated correct winch.

7. Crew should attend the live end of the winch.

8. Crew should keep designated tie off points for which cable not in use. This may be at the base of winch, near the V door or other Location as specified by winch operating procedure.

59. Pulling and installing master and inner Bushing

1. Crew should Use PPE

2. Remove the inner bushing first then remove the master bushing

3. Master Bushing to be raised slowly and kept aside.

4. Rotary hole to be greased properly and muster bushing to be placed

5. After Master bushing placing grease it and place inner bushing

6. Rotary hole should be covered with hole cover

60. Drilling Fluid Control

1. Crew should use proper PPE

2. Efficiently removals of solid from drilling fluid help ensure highest quality of drilling fluid.

3. Improved solid control procedure includes less replacement fluid, fewer additives, loss waste.

4. Good fluid control causes less risk of hole problem related to excessive solids content.

5. It reduces drilling fluid cost

6. Fluid control crew should undergo training.

7. Crew should use Rigorous integrated operational and technical process.

8. Highest quality of solid control equipment to be maintained.

61. Displacing hole with OBM

1. Crew should use proper PPE

2. In case of loss circulation pill to be given and loss to be controlled.

3. Sufficient amount of oil base mud is to be prepared.

4. Solid Control Engineer to be sent to rig for fluid properties control.

5. Solid control equipment to be provided at rig.

6. OBM is to be used for particular formation as shale for inhibition.

7. If drilled with OBM floor area and working area should be clean.

8. Any spillage of OBM to be stopped.

9. In OBM drilling rate of penetration will be fast and should not be confused with drilling break.

10 If gas kick cones it may be dissolved in OBM and migrate quickly.

62. Displacing OBM to sea water & Vice versa

1. Crew should use PPE

2. Crew should clean all surface equipment.

3. Crew should check fluid flow path to be clear by circulation.

4. Crew should calculate pressure differential along the flow path to reduce over pressuring casing or tubing.

5.. Spacer densities be carefully designed to minimize pressure differential.

6. Run and Scrap upto casing bottom.

7. Do not stop pumping at any time during displacement until continuous flow of clear brine fluid.

8. Drilling fluid should not be mixed with Zink.

9. Pumping pressure to be checked continuously because brine reacts with OBM and when they are mixed generally they produce viscous

unpumpable mass due to flocculation of mud by high salt content of brine should these happen pumping to be stopped.

10. It is required to remove Sulpher or Sulphide from contamination for corrosion control.

63. Back loading OBM.

1. A gas test for LEL and H^2S should always be tested performed on the tank before offloading.

2. MSDS sheet should be filled up.

3. Back load master to check dirty tank.

64. Cleaning mud Pit.

1. Crew should use proper PPE.

2. No personnel exposure to hazardous material.

3. No slip, trip or falls in what inherently or slippery environment.

4. No personnel entrapment in confined space.

5. No personnel exposure to fire and explosion in a confirmed space.

65. Entry in Mud Pit.

1. Crew should use proper PPE.

2. Restrict access to unauthorized personnel.

3. Impose confirmed space entry procedure for entry in mud pit.

4. Keep access door closed.

5. Use low toxic base oil.

6. Provide fixed alarm for hydro carbon and H2s.

7. Ensure that samples follow the rules for access to the area.

66. Hot Work in Mud Pit.

1. Crew should use proper PPE.

2. Perform hot work in safe location or with fire hazard removed or covered.

3. Use guards to continuous heat, sparks to protect immovable fire hazard.

4. Make suitable fire extinguisher, bucket of sand readily available.

5. Be familiar with sounding of alarm.

6. Make fire watch atleast half hour after hot work.

7. Inspect welding and cutting unit.

8. Review hot work permit.

67. Chemical mixing.

1. Use proper PPE.

2. Keep Eye Wash near mixing room.

3. The health effects include dizziness, headaches, drowsiness, nausea as well as dermatitis and causing irritation and inflammation of the respiratory system and cancer.

4. Restrict access to unauthorized personnel.

5. Use bulk transfer method where possible.

6. Enclose the flopper as much as possible and provide local exhaust ventilation. Permit access for careful opening and emptying sacks.

7. Ensure solids are poured gently.

8. Ensure empty sack are rolled up in the extracted zone and put in a polythene sack.

9. Fit an airflow indicator to show that extraction is working properly.

10. Discharge extracted air to safe place.

68. Handling of Chemical.

Same As Above.

69. Handling of toxic or Hazardous Chemicals.

1. Correct labelling of containers and pipework, using warning placards and outer warning placards and displaying of safety signs
2. Maintaining a register and manifest (where relevant) of hazardous chemicals and providing notification to the regulator of manifest quantities if required
3. Identifying risk of physical or chemical reaction of hazardous chemicals and ensuring the stability of hazardous chemicals
   .Ensuring that exposure standards are not exceeded
4. .Provision of health monitoring to workers
5. .Provision of information, training, instruction and supervision to workers
6. .Provision of spill containment system for hazardous chemicals if necessary
7. .Obtaining the current Safety Data Sheet (SDS) from the manufacturer, importer or supplier of the chemical
8. .Controlling ignition sources and accumulation of flammable and combustible substances.

9. .Provision and availability of fire protection, fire fighting equipment and emergency and safety equipment
10. Preparing an emergency plan if the quantity of a class of hazardous chemical at a workplace exceeds the manifest quantity for that hazardous chemical
11. .Stability and support of containers for bulk hazardous chemicals including pipework and attachments
12. Decommissioning of underground storage and handling systems
13. Notifying the regulator as soon as practicable of abandoned tanks in certain circumstances
14. The WHS Regulations contain prohibitions or restrictions on certain hazardous

70. Mixing Caustic Soda.

Operators that handle caustic soda should be required to observe the operating
standard for safe operations. For this, it is necessary to provide education
and training concerning:
1. The characteristics, level of hazard, and methods of handling of caustic
     soda
2. The location of protectors, showers, eye washers, water taps, cleaning
   Hoses and first aid facilities
3. Proper method for the use of protectors and first aid facilities
4. First aid measures to be taken in case of an emergency
5. For operators filling tanks, measures for preventing a lack of oxygen.
6. It is also important to train supervisor concerning the following, and regularly
   carry out training drills for dealing with disasters:
7. Proper usage of the first aid facilities
8. Measures to be taken in the case of a chemical injury.

9. (2) Operational rules
10. It is important to establish rules concerning the proper use the facilities for
handling caustic soda or any associated facilities, and to operate them in accordance
with the rules.
11. (3) Voluntary inspection
12. Caustic soda is a highly corrosive substance. It is important to periodically
inspect equipment that is used for handling caustic soda and to retain the
inspection records.

Prevention Measures for Hygiene

13. The storage of food in places where caustic soda is stored or working areas
14. where it is handled is prohibited, and smoking, eating or drinking in such
places is also prohibited.
15. The operator must wear cotton or synthetic-fiber work wear, a work cap, protective
goggles, rubber boots, rubber gloves, and a rubber apron. In the area
where mist or dust is in the air, wear a dust mask.
16. Protective cream has no effect. Wash the face and hands thoroughly at the
end of the work operations.
17. In places where caustic soda is handled, provide a shower and face washing
facility that can be used immediately at any time

71. CHANGING THE ELEVATOR LINK

1. Use proper PPE
2. Lower the T/Block slowly and hold the link.
3. After the Block is lowered hang Link with air winch.

4. Take out Link from Travelling Block
5. Check the Link rating for string. Use proper Link depending on suitability.

72. CHANGING THE ELEVATOR

1. Use proper PPE
2. Lower the T/Block so that Elevator Rest on ground.
3. Open the elevator handle and shift the elevator.
4. Keep the elevator and push Link in the arm of elevator.
5. Tighten the screw of Elevator arm.

73. RIG UP WIRELINE SHEAVE

1. Use proper PPE
2. Connect Air tugger To Sheave.
3. Pass wire line through sheave and lift the sheave.

74. JARRING OPERATION

1. Use proper PPE
2. While jarring operation pull jar slowly on drill string.
3. Stretch the drill string.
4. After stretching jar reaches firing point.
5. Jar gives an upward or downward impact load for releasing string.
6. Jar should be above drill collar.
7. Drill collar should not be used above jar.

8. Service the jar properly.
9. Jar to be set while lowering.

75. POST JARRING OPERATION

1. Use proper PPE
2. After jarring and release of string circulate and keep string free.
3. Check shaker for cutting.
4. Pull out of hole.
5. Check and watch for drill pipe damage while pulling out.
6. Lay down jar and clean service it.

76. PERFORATING

1. Crew should use proper PPE.
2. Pick up wire line sheave and shooting nipple.
3. Check perforation gun.
4. While loading gun crew should stay away from vicinity.
5. Keep Radio silence and announce .
6. Use proper charge as per perforation plan.
7. After pulling out of perforating gun check for any part of tool left in hole.
8. If radio actives are left in hole separate contingency – plan is to be followed.

9. During pulling out and 100m near Rotary Radio Silence to be observed.
10. Crew should stay away when perforation gun on Rig floor.

77 RADIOSILENCE

1. Announce Radio Silence
2. Inform other locations, platform for Radio silence.
3. Close and turn off radio.
4. Radio silence for explosives and perforation job.
5. Announce after radio silence.

78. EXPLOSIVE

1. Use proper PPE.
2. Keep explosive in separate area in the Rig with level explosive.
3. Entry to be restricted to authorized personnel.
4. Carrying of explosive to be maintained as per procedure.
5. In Rig explosive manuals to be kept.
6. Manifest of explosives and data sheet to be maintained.
7. Explosive Room should be under Lock and Key.

79. WIRELINE

1. Use proper PPE.
2. Wire Line to be checked for damage, pinch point and kink and rust.

3. Damages wire line should be cut away.

80. TRIP WITH WIRE LINE
1. Use proper PPE.
2. Do not touch Wire Line when tool is getting lowered.

81. WELL TESTING
1. Use proper PPE.
2. Keep the well testing plan.
3. Test as per procedure.
4. Check testing equipment.
5. Check all the testing lines.
6. Proper reporting and data entry for testing to be recorded.

82. INSTALLING / REMOVING BURNER BOOM
1. Use proper PPE.
2. Install Burner Boom vertically and connect to Rig.
3. Use pin to connect Burner Boom.
4. Lower Burner Boom and secure with cable on End and fasten with Rig.

83. LOWERING BURNER BOOM
1. Use proper PPE
2. Lower Burner Boom with crane.
3. Connect cable to Burner Boom.

84. RAISING THE BURNER BOOM
1. Use proper PPE
2. Use crane for lifting the boom.

85. RIG UP HOSES FOR BURNER BOOM
1. Use proper PPE
2. Hoses to be connected to Burner Boom.
3. Use hammer union to connect hoses.
4. While hammering restricted area caution should be taken.

86. FLARING
1. Use proper PPE
2. Close the well from burner and main choke manifold.
3. One person should go to the end of burner and place a burning rack in diesel and place near burner.
4. Open the well for flaring.

87. MAKE UP SHOE AND FLOAT COLLAR ON DECK
1. Use proper PPE
2. Keep easing pipe on main deck.
3. Clean thread properly.
4. Use backer lock thread compound.
5. Connect float shoe and collar.
6. Tighten with chain tong.

88. CHANGING THE DRILLING LINE

1. Use proper PPE
2. Hang the block.
3. Loosen the dead end line.
4. Wrap the casing line on main drum.
5. Connect new easing line spool.
6. Mark the cable and open the line.
7. Cut and remove the old line.
8. Connect new line end to easing drum.
9. Tighten the dead end nut.

89. INSTALLING THE HANDRAIL

1. Use proper PPE.
2. Prepare hand rail by holding.
3. Place hand rail on socket and weld.
4. Reweld socket with hand rail.

90. SCRAPPING AND CHIPPING

1. Use proper PPE.
2. Use hand grinder for chipping operation.
3. Always use safety google.

91. PICKING UP TUBULUR FROM BOAT

1. Use proper PPE

2. Deckman should give signal to crane operator and vice versa.
3. Crew should use walkie talkie to talk to boat.
4. While lowering the boom boat crew should give signal.
5. Boat captain should supervise the boat crew operation.
6. Proper sling should be used.
7. Crane operator should lift load cautiously and smoothly and not by jerk.

92. LOADING / BACKLOADING FROM / TO THE BOAT

1. Use proper PPE.
2. Crew should bunch the tubular by a good string.
3. There should be proper signaling between Rig crew and crane operation.
4. Crane operator should lift the load and place on boat's main deck.
5. Craw should check for any loose item.
6. Slings should be of correct capacity for moving, picking or lowering.
7. While loading nobody should stand below load.

93. SUPPLY / RETRIEVING TUBULAR OR ANY OTHER MATERIAL FROM THE RIG FLOOR

1. Use proper PPE
2. Pick up the material by using crane.

3. Crew should give proper signals.
4. Crane operator should lift the material carefully/
5. Crew should not lift material by nylon rope.

94. GENERAL CLEANING

1. Use proper PPE
2. Remove garbage from cleaning area.
3. Use bucket with soap water and clean with brush.
4. After cleaning with soap water thoroughly clean with drill water.

95. MANUAL LIFTING AND MATERIAL HANDLING

1. Use proper PPE
2. Use proper stance.
3. Lift the load properly without harming yourself.
4. Always carry load by two persons.

96. USING CHAIN BLOCK / TROLLEY HOIST FOR LIFTING

1. Use proper PPE.
2. Chain block to be checked properly for pulling chain freely.
3. Beam for hoisting chain blocked to be checked for support of load.
4. Chain block is to be operated smoothly.

100. PERSONNEL TRANSFER FROM / TO BOAT

1. Crew should use proper PPE
2. Personnel should put on life jacket.
3. Crew should stand on rim of basket and hold rope.
4. Crane operator should lift basket properly.
5. Deck crew should give proper signal.
6. Walkie talkie to be used by crew.
7. Crane operator should lower the basket to boat.

101. USING WORK BASKET

1. Crew should use proper PPE.
2. Crew should check basket for any damage.
3. Slings for hoisting to be checked.
4. Safety harness to be provided during work in basket.
5. One crew should be specially assigned for signaling.
6. When winch is not in use it should be secured.
7. Display of poster should be kept on winch.
8. Communication should be done by walkie talkie.

102. WORKING OVERBOARD

1. Crew should use proper PPE
2. Safety belt to be used.
3. When working on scaffolding competency certificate is required.

4. Work should not be carried out in the night.

103. USING CAT HEAD

1. Crew should use proper PPE.
2. Extra care to be taken during tightening so that hands and fingers do not crush between cat head and manila rope.

104. USING CHAIN TONG

1. Crew should use proper PPE.
2. Check that chain tong when in use does not hurt any body part.
3. When not in use chain tong should be clean and kept in diesel oil drum.
4. When tightening chain tong should be checked for slippage.

105. USING PIPE WRENCH

1. Crew should use proper PPE.
2. Use proper size pipe wrench.
3. Keep tool clean.
4. When tightening with pipe wrench watch for enough space for movement.

106. USING HAND TOOL

1. Crew should use proper PPE.
2. Clean the tool by diesel and keep in box.

107. STORAGE HANDLING, MAINTENANCE AND USE OF LIFTING SUB AND LIFTING CUPS

1. Crew should use proper PPE.
2. Thread end should be cleaned.
3. Lifting sub to be fitted with lifting cups.
4. Lifting sub should be lifted by overhead crane.
5. NDT inspection should be carried out of lifting sub.

108. RABBITING THE CASING ON DECK

1. Crew should use proper PPE.
2. Keep the casing on pipe rack.
3. Do not keep foot on casing pipe.
4. Roll casing and use crowbar.
5. Rabbit casing.
6. Clean the garbage inside casing.

109. USE OF CASING THREAD PROTECTOR

1. Crew should use proper PPE.
2. Clean the casing thread with diesel.
3. Apply casing dope.
4. Put on casing thread protector.
5. Avoid cross threading while putting protector.

110. RABBITING THE DP / HWDP / DC ON THE DERRICK

1. Use proper PPE.
2. Drop the rabbit on top of monkey board.
3. Do rabbiting during RI.

112. RABBITING THE DP / HWDP / DC ON THE V DOOR

1. Use proper PPE
2. Drop the rabbit on box end.
3. Check that no one should be on catwalk.

113. CLIMBING MONKEY LADDER

1. Crew should use proper PPE.
2. Use safety belt.
3. Climb the monkey ladder.

114. USE, CARE AND MAINTENANCE OF LADDER SAFETY SYSTEM

1. Crew should use proper PPE.
2. Safety system should be checked for wear and tear.
3. Rope to be checked.

115. USE OF STAIRWAY

1. Crew should use proper PPE.

2. While using stairway look at stairway and walk.

3. Watch the handrail.

116. RIG UP COIL TBG UNIT

1. Use proper PPE

2. Pick up CTU unit by crane on rig.

3. Pick up CTU unit by block.

4. Make up chikson connection with equipment.

.117. RIG UP CHICKSON LINES

1. Use proper PPE

2. Connect chikson line with hammer union.

118. INSTALL CIRCULATING SUB ON SLING AND RIG UP CONNECTION HOSE

1. Use proper PPE

2. Use circulating sub on sling and tighten it.

3. Pick up   hose.

4. Connect hammer union with hose on circulating sub.

119. RIG UP RIG DOWN HYDRAULIC HOSE.

1. Use proper PPE

2. Connect hydraulic hose by hammer union.

3. Initially hand tight hammer union.

4. Tighten hammer union by hammering.

120. NIPPLE UP X-MAS TREE

1. Use proper PPE
2. After running in compleation sting land tbg hanger.
3. Remove BOP
4. Pick up X- MAS tree.
5. Check all valve opening.
6. Place X-MAS tree on well head.
7. Make alignment of X-MASS tree.

121. STRIPPING IN HOLE.

1. Use proper PPE.
2. Use maximum pressure for stripping on annular is equal to 40% of the maximum static pressure rating dry or 60% if the pipe is lubricated as it is being stripped through the annular.
3. While pulling out check stripping ram opening to avoid fishing operation due to pipe damage.

122. Installation, Operation and maintenance of Hydraulic Elevator

1. Crew should use proper PPE
2. Crew should check opening or closing of Hydraulic Elevator.
3. Check for wear and tear of hoses, if warn out replace them.
4. Check the Hydraulic oil level.
5. Grease the elevator.
6. Close and open the elevator until the operation is smooth.
7. Grinding or Welding is not allowed.
8. Check for elevator sway from hitting.
9. Use elevator recommended spare part.

123. Installation, operation and maintenance of manual elevator.

1. Use proper PPE
2. Check the opening and closing of elevator.
3. Check the elevator sway so that it does not hit the body part.
4. Grease the elevator.
5. While opening the elevator care should be taken not to touch any place other than handle opening.
6. Grease the elevator.
7. Check for pin or any other loose part in elevator.
8. Use elevator recommended spare part.

124. Installation, Operation and maintenance of casing elevator

1. Use proper PPE.

2. Use proper rated casing elevator for casing pipe weight.

3. Watch for loose part in elevator.

4. Lift casing elevator shortly.

5. Grease casing elevator.

6. Slack elevator while making or Breaking pipe.

7. Watch for pin are put in place.

8. Use elevator recommended spare.

125. Use of single JT elevator

1. Use proper PPE.

2. When putting single jt elevator two people should hold it.

3. Pin of elevator to be checked properly after latching.

4. Slings for elevator to be checked.

5. Use elevator recommended spare.

126. STANDARD OPERATING PROCEDURE

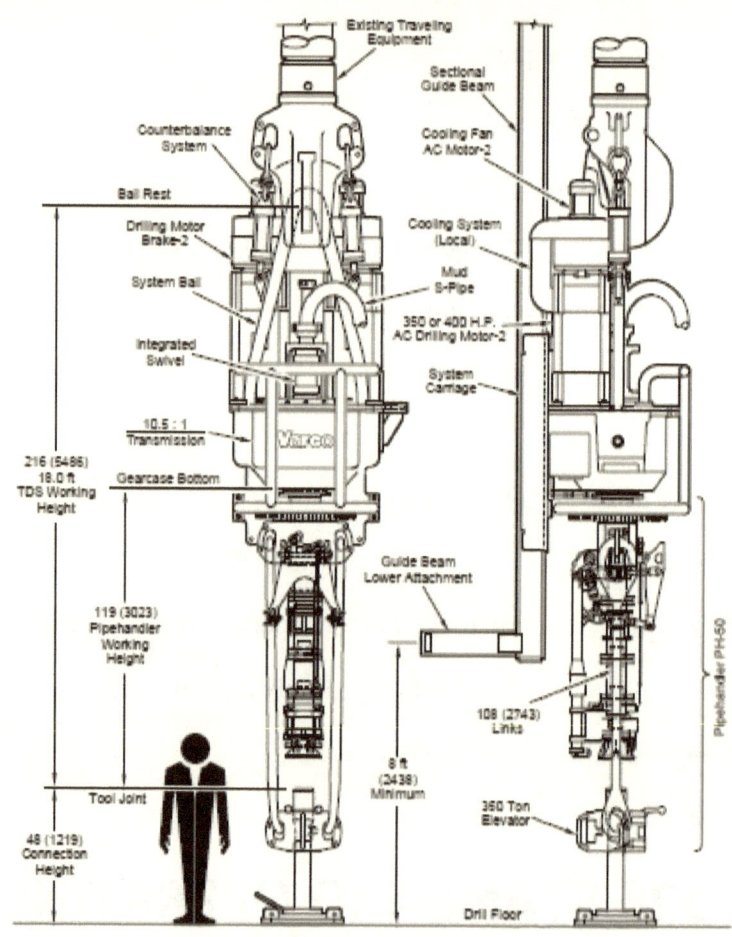

TDS-11SA General Arrangement

1320 UDBE DRAWWORKS OPERATION

! ATTENTION !

Before performing any service function, be certain that the unit is separated from the power source or that the power source is locked-out to prevent any form of energy from entering the equipment. This would include electrical or mechanical energy, into or from the prime/mover(s); pneumatic energy from the compressor/air system, etc. The travelling equipment should be suspended from the hanging line and the mechanical brake released to insure there is no load on the drum that may cause unintended rotation. The sub-assemblies of the unit represent massive weights and must also be suspended being released from operating.

WARNING
Failure to observe the **WARNINGS** can result in **property damage, serious bodily injury or even death!**

Specifications

The Type 1320-UE Drawworks is rated for 2000 combined engine horsepower. Within the designed capacity of the drawworks, the maximum loads that should be handled are shown in the table below:

Clutch	Low		High	
Transmission	Low	High	Low	High

Load in thousands Of pounds	8 lines	660	415	270	170
	10 Lines	805	505	330	205
	12 Lines	940	590	380	240

- **Nominal drilling depth range 13,000 to 20,000 ft.**
- **Wireline Size 1-3/8"**

Chain Specifications

1-1/2" Sextuple Chain (ASA-120-6)
Transmission Low Speed
78 Pitches
Transmission High Speed
68 Pitches
2" Quadruple Chain (ASA-160-4)
Low Drum Drive
122 Pitches
High Drum Drive
102 Pitches
2" Double Chain (ASA-120-2)
Rotary Countershaft Drive
82 Pitches
Catshaft Drive
88 Pitches
Weights (Estimated)
Drawworks Front Section (less optional equipment)
41,660#
Core Reel
4,300#
60" Hydromatic Brake with Cradle and Type "A" Over-running Clutch   10,890#

Drawworks Rear Section
31,700#
Drawworks Complete as Above
88,550#

Drawworks Dimensions

Width, Front Section
        7'-9-3/8"
Width, Rear Section
        6'-2-13/16"
Width, Overall
        13'-11-9/16"
Length, Overall
25'-0"
Height, with Core Reel
9'-6"
Height, less Core Reel
9'-4"

127. STANDARD OPERATING PROCEDURE

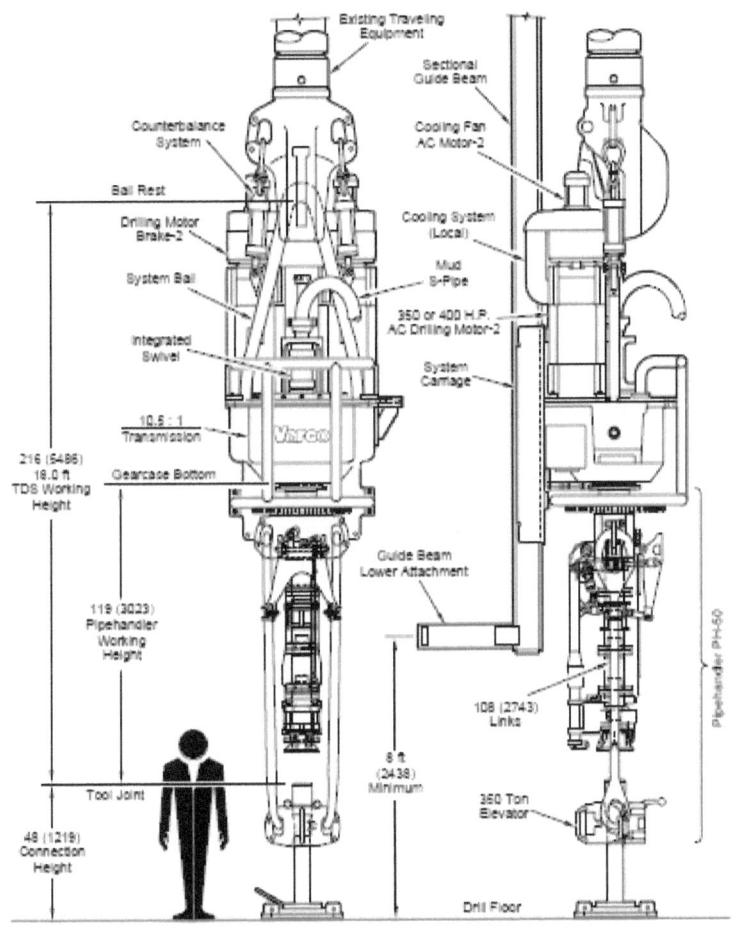

TDS-11SA General Arrangement

TDS - 11 SA TOP DRIVE SYSTEM.

STANDARD OPERATING PROCEDURE
TDS 11 SA TOP DRIVE SYSTEM
PPE REQUIREMENT: RIG FLOOR PPE.

Purpose: TDS 11 SA operation.

Initial condition: The TDS is installed on the mast, VFD is functional, all electrical connections are complete and functional, All hydraulic functions are checked.

Reference documents: TDS- 11SA Top Drive System operations manual.

Job hazards: As per the rig hazard analysis manual.

Tools required: None.

Operations:

1. Breaking out of Drill Pipe from TDS-11SA

Steps	Key points
1. Stop Rotation.	

2. **Shut down** Mud Pump. 3. **Close** TDS-11SA IBOP. 4. Select **Reverse, Off, Forward Switch** to **REVERSE position**. 5. **Push and HOLD** onto the **Torque Wrench Button** and Check Shot Pin is Engaged. (Go to **Note** if Shot Pin does not engage, **DO NOT** attempt to **Rotate** the **Rotating Head!!**). 6. Select **Drill, Spin, Torque Switch** to **Torque and Hold** till Saver Sub Break out from Drill pipe. (Observe the reading on the Torque Meter as the break out torque maybe higher than pre-mark up torque due to down hole torque). 7. Select **Drill, Spin, Torque Switch** to **Spin position**. 8. Select **Stand Jump Switch** to **Stand Jump position**. (Beware as TDS-11SA will rise itself in this mode) 9. Select **Stand Jump Switch** to **Drill position**. (Beware as TDS-11SA will lower itself in this mode). 10. Select **Drill, Spin, Torque Switch** to **Drill Position**.	If Shot Pin did not Engage when Torque Wrench is Pushed and Hold, Release the Torque Wrench Button. Rotate the Rotating Head either right or left slightly. Repeat Step 5 onward as above.

2. Making Drill Pipe onto TDS-11SA

Steps	Key points
1) Stab the Drill Pipe onto the existing pipe on the drill floor. 2) **Secure Rig Back Tong** onto the **existing pipe** on drill floor. 3) Select **Spin mode** on the **Drill, Spin, Torque Switch**. 4) Slowly lower TDS-11SA. 5) Once Drill Pipe is fully threaded onto TDS-11SA Saver Sub and existing pipe on drill floor, Select **Torque mode** on the **Drill, Spin, Torque Switch**. 6) **Observe** the reading on the **Torque Meter** to the **Set Torque value**. 7) Select **Drill mode** on the **Drill, Spin, Torque switch**.	

3. Setting of Drilling Current/Torque

Steps	Key points
1) Select **Brake** to **ON position**.	If Shot Pin did not Engage when Torque Wrench is Pushed and Hold, Release the Torque Wrench Button. Rotate the Rotating Head either right or left slightly. Repeat Step 5 onward as above.
2) **Open** the **Throttle** to ¼ turn.	
3) **Adjust** the **Drill Limit Current Knob** to the require Drill Torque by observing the reading on the Torque Meter.	
4) Once the require Drill Torque is adjusted, **Close** the **Throttle**.	
5) Select **Brake** to **OFF position**.	

4. Setting of Making Up Current/Torque

Steps	Key points
	If Shot Pin did not Engage

1)	Set **Brake** to **ON** position.	when Torque Wrench is Pushed and Hold, Release the Torque Wrench Button. Rotate the Rotating Head either right or left slightly. Repeat Step 5 onward as above.
2)	Select **Torque mode** and **HOLD** on the **Drill, Spin, Torque Switch.**	
3)	**Adjust** the **Make Up Current Knob** to the require Torque by observing the reading on the Torque Meter.	
4)	Once require Torque is set, select **Drill mode** on the **Drill, Spin, Torque Switch**.	
5)	Select **Brake** to **OFF** position.	

5. Drilling ahead with triples

Steps	Key points
a. Drill down the existing	If Shot Pin did not Engage when Torque Wrench is

stand and set the slips. b. Breakout the saver sub from the drill pipe using the top drive motor and backup clamp in the pipehandler. c. Spinout the connection using the drilling motor. d. Lift the top drive. e. The derrickman latches the triple in the elevator and the floor crew stabs it in the box. f. Lower the top drive, stabbing the pipe into the stabbing guide until the pin of the saver sub enters the box. g. Spin-up and torque the connection using the drilling motor (make-up torque must be preset). Use a backup tong to react the torque. h. Pull the slips, start the mud pumps and drill ahead.	Pushed and Hold, Release the Torque Wrench Button. Rotate the Rotating Head either right or left slightly. Repeat Step 5 onward as above.

6. Drilling ahead with singles

Steps	Key points
a. Drill down the existing joint and set the slips. b. Breakout the saver sub from the drill pipe using the top drive motor and the backup clamp in the pipehandler. c. Spin out the connection using the drilling motor. d. Lift the top drive until elevators clear box. e. Actuate the link tilt to bring the elevator over to the single in the mousehole, lower the top drive and latch	If Shot Pin did not Engage when Torque Wrench is Pushed and Hold, Release the Torque Wrench Button. Rotate the Rotating Head either right or left slightly. Repeat Step 5 onward as above.

the elevator around the single in the mousehole.

f. Pull the single out of the mousehole and as the pin clears the floor, release the link tilt to allow the single to come to well center.

g. Stab the connection at the floor and lower the top drive allowing the added single to enter the stabbing guide.

h. Spin-up and torque the connection using the drilling motor (torque mode). Set a backup tong to react the torque.

i. Pull the slips, start the mud pumps and drill ahead.

7. Tripping in and tripping out

Tripping is handled in the conventional manner. The link tilt feature can be used to tilt the elevator to the derrickman, enhancing his ability to latch it around the pipe thus improving trip times.

The link tilt has an intermediate stop which is adjustable to set the elevator at a convenient working distance from the monkey board. The intermediate stop is tilted out of the way to allow the elevator to reach the mousehole.

The elevator may rotate in any direction from frictional or torque forces realized by the drill string.

If a tight spot or key seat is encountered while tripping out of the hole, the drilling motor may be spun into the stand at any height in the derrick and circulation and rotation established immediately to work the pipe through the tight spot.

8. Back reaming

Procedures for reaming out of the hole:

a. Hoist the block while circulating and rotating the string until the third connection appears.

b. Stop circulation and rotation, and set the slips.

c. Breakout the stand at floor level, and spin out using the drilling motor.

d. Breakout the drilling motor from the top of the stand using the top drive motor and backup clamp, then spin out with the drilling motor.

e. Pick up the stand with the drill pipe elevator.

f. Rack stand back.

g. Lower the top drive to the floor.

h. Stab drilling motor into box, spin-up and torque with the drilling motor. With light slip loads, the top drive and backup clamps can be used to torque the connection.

i. Resume circulation and continue reaming out of the hole.

9. Breaking out the saver sub

During normal operation, the torque backup clamp cylinder is sitting on the springs, which are supported by the bottom plate of the torque arrestor.

1. Loosen the tool joint lock between the saver sub and the lower IBOP valve by unscrewing the ten bolts. Refer to the tool joint lock assembly and disassembly procedures in the Maintenance section. Slide the tool joint lock down until it rests on the clamp cylinder body.

2. Raise the clamp cylinder until the clamp cylinder positioning slot lines up with the first hole on the torque arrestor. Insert the safety pin through the clamp cylinder and torque arrestor.

3. Select TORQUE mode. Pressurize the clamp cylinder to clamp on the saver sub by pressing and holding the TORQUE WRENCH PRESS AND HOLD button.

4. Switch the drilling motor to REVERSE to break out the connection.

5. Once the connection is broken out, switch to SPIN and allow the motor to spin until the saver sub and lower IBOP valve separate. Remove the safety pin. Lower the clamp cylinder with the saver sub. The saver sub is ready for removal.

6. Unclamp the saver sub by releasing the TORQUE WRENCH PRESS AND HOLD button.

Stand clear. The saver sub must be supported before unclamping it. It will fall through the bottom of the stabbing guide if not supported.

10. Making up the saver sub

1. Manually screw in the replacement saver sub into the lower IBOP valve. To manually screw in the replacement saver sub to the lower IBOP valve, raise the clamp cylinder until the lower IBOP valve is exposed below the stabbing guide (a pup joint may be used). Lower the clamp cylinder using the winch until the hole and slot line up. Insert the pin.

2. Select TORQUE mode. Pressurize the clamp cylinder to clamp on the saver sub by pressing and holding the TORQUE WRENCH PRESS AND HOLD button.

3. Switch the drilling motor to FORWARD. Select SPIN mode and rotate the drilling motor until the saver sub shoulders against the lower IBOP valve. Select TORQUE mode and apply the desired torque.

4. Release the TORQUE WRENCH PRESS AND HOLD button to unclamp. Lower the clamp cylinder all the way down.

5. Position the tool joint lock correctly and follow the proper assembly procedure described in the Tool joint locks section.

11. Breaking out the lower IBOP

Remove the saver sub.

1. Loosen the tool joint lock between the lower IBOP valve and the upper IBOP valve by unscrewing the bolts. Slide it down and rest it on the tool joint lock sitting on the clamp cylinder.

2. Raise the clamp cylinder with the two tool joint locks until the clamp cylinder slot lines up with the second hole on the torque arrestor. Insert the pin.

3. Select TORQUE mode. Pressurize the torque backup clamp cylinder to clamp on the lower IBOP valve by pressing and holding the TORQUE WRENCH PRESS AND HOLD button.

4. Once the connection is broken out, switch to SPIN and allow the motor to spin until the upper IBOP valve and the lower IBOP valve separate.

5. Remove the safety pin. Lower the clamp cylinder with the lower IBOP. The lower IBOP is ready for removal.

6. Unclamp the lower IBOP valve by releasing the TORQUE WRENCH PRESS AND HOLD button.

Stand clear. The lower IBOP valve and saver sub must be supported before unclamping them. They will fall through the bottom of the stabbing guide if not supported.

12. Making up the lower IBOP

1. Screw in the replacement saver sub and the lower IBOP valve together manually and stand them under the clamp cylinder (a pup joint may be used to support it). Position the clamp cylinder by stabbing over the lower IBOP valve. Make sure the lower IBOP valve comes up through both tool joint locks sitting on the clamp cylinder body. Tighten four alternate screws on the top tool joint lock to secure it to the lower IBOP valve to provide a temporary shoulder to support the weight of the lower IBOP valve and the saver sub.

Make sure all four screws are tightened sufficiently so that the tool joint will not slide through when the clamp cylinder is raised.

2. Select SPIN and FORWARD modes.

3. Raise the clamp cylinder with the lower IBOP valve and saver sub while rotating the upper IBOP to engage the threads. Once the upper IBOP valve and the lower IBOP valve start to spin together, stop the drilling motor.

4. Lower the clamp cylinder and line up the first slot and hole on the clamp cylinder and the torque arrestor. The clamp cylinder jaws line up with the saver sub.

5. Select TORQUE mode. Pressurize the torque backup clamp cylinder to clamp on the saver sub by pressing and holding the TORQUE WRENCH PRESS AND HOLD button.

6. Switch the drilling motor to FORWARD. Select SPIN mode and rotate the drilling motor. Select TORQUE mode and apply desired torque and makeup both connections.

7. Release the TORQUE WRENCH PRESS AND HOLD button to unclamp. Lower the clamp cylinder all the way using the winch.

8. Loosen the temporarily made-up tool joint lock. Position both tool joint locks correctly and follow the proper assembly procedure described in the Tool joint locks section.

13. Breaking out the upper IBOP

Refer to the illustration on the next page. Remove the saver sub and the lower IBOP first.

The saver sub and the lower IBOP can be removed as one unit by breaking the connection between the upper and lower IBOPs.

1. Lower the clamp cylinder with the broken out lower IBOP valve and the saver sub.

2. Unclamp the lower IBOP valve/saver sub assembly by releasing the TORQUE WRENCH PRESS AND HOLD button.

3. Remove the two tool joint locks sitting on the clamp cylinder.

Stand clear. The lower IBOP valve and saver sub must be supported before unclamping them. They will fall through the bottom of the stabbing guide if not supported.

4. Remove the IBOP actuator yoke by unpinning it at three places.

5. Remove the two upper IBOP cranks by unscrewing the 2 sets of screws.

6. The IBOP actuator shell stays on the upper IBOP valve assembly.

7. Loosen the top tool joint lock and rest it on the actuator shell.

8. Raise the clamp cylinder with the actuator shell and the tool joint lock until the third slot and hole line up. Insert the pin.

9. Select TORQUE mode. Pressurize the clamp cylinder to clamp on the upper IBOP valve by pressing and holding the TORQUE WRENCH PRESS AND HOLD button and switch to REVERSE.

10. Switch the drilling motor to REVERSE to break the connection.

11. Once the connection is broken out, switch to SPIN and allow the motor to spin until the upper IBOP valve and drive stem separate.

12. Remove the safety pin. Lower the clamp cylinder with the upper IBOP. The upper IBOP is ready for removal from the clamp cylinder.

13. Unclamp the upper IBOP valve by releasing the TORQUE WRENCH PRESS AND HOLD button.

14. Remove the tool joint lock and the actuator shell.

14. Making up the upper IBOP

1. Place the upper IBOP valve on the floor under the clamp cylinder so that the clamp cylinder can be stabbed over it (a pup joint may be used to support it). Lower the clamp cylinder so that the upper IBOP comes up through the clamp cylinder. Place the actuator shell and tool joint lock over the upper IBOP and tighten four alternate locking screws to secure it to the upper IBOP, providing a temporary shoulder to support its weight.

2. Install the actuator shell and the cranks.

3. Raise the clamp cylinder with the upper IBOP valve while rotating the drive stem clockwise to engage the threads. Once the drive stem and upper IBOP valve start to spin together, stop the motor, switch to FORWARD and SPIN and press and hold the TORQUE WRENCH PRESS AND HOLD button to spin in.

4. Spin the lower IBOP and saver sub into position (refer to the appropriate procedures in the previous sections). Make sure that the two tool joint locks are properly installed on the clamp cylinder and in the correct sequence.

5. Release the TORQUE WRENCH PRESS AND HOLD button to unclamp. Lower the clamp cylinder until it lines up with the first hole. Now the clamp cylinder is lined up with the saver sub.

6. Select TORQUE mode. Pressurize the clamp cylinder to clamp on the saver sub by pressing and holding the TORQUEWRENCH PRESS AND HOLD button and apply desired torque to makeup all three connections.

7. Place the three tool joint locks at their respective joints. Install the three tool joint locks by using the proper assembly procedure described in the Tool joint locks section.

8. Install the IBOP actuator yoke and secure it.

9. Remove the winch and store it in its box.

15. CASING LINE CUT / SHIFT

1. Lower the TDS to Rig Floor.
2. Locate the RIG-UP / SHUT DOWN / RUN valve on the hydraulic manifold as shown in the figure.
3. With the hydraulic power ON select the "RIG-UP" position and remove the extended counter balance cylinders from the hook ears .

4. Select "SHUT DOWN" position and turn off the top drive.
5. Isolate all power to Varco control house.
6. Engage the upper and lower carriage locks as shown in the figure.
7. Install bail lock with the goose neck.
8. Remove the bail from the travelling block hook.
9. Raise the travelling block to safe height.
10. Carry out slip / cut as given below:
    For slip.
    - Apply the brake on the dead end anchor.
    - Loosen the dead end anchor bolts.
    - Slip the required length of casing line.
    - Tighten the dead end anchor bolts.

    For Slip and cut.
    - Suspend the travelling block with casing line from the eye pad on rig floor with the help of sling and clamp.
    - Remove the fast end wire line clamp assembly.
    - Cut and remove required length of casing line.
    - Refit the fast end wire line clamp.
    - Slip the required length of casing line as per the procedure mentioned above.
11. Engage the travelling block hook with the top drive bail.
12. Remove the bail lock.
13. Switch on the power to the varco control house.
14. With the hydraulic power ON select the "RIG-UP" position and connect the cylinder clevis to the pear link.

128. STANDARD OPERATING PROCEDURE

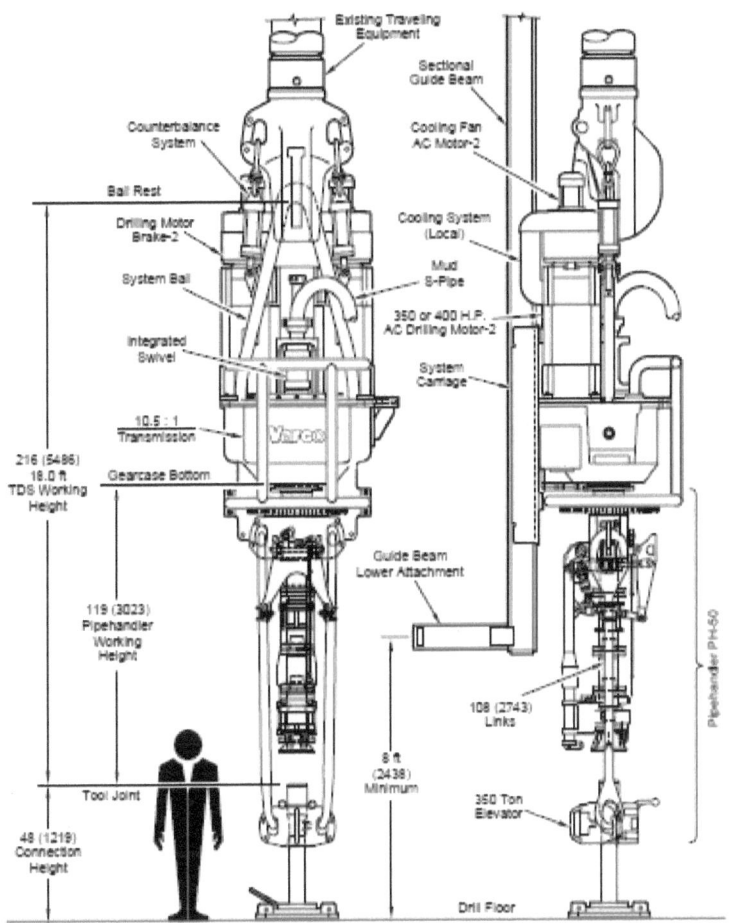

TDS-11SA General Arrangement

12-P-160 MUD PUMP OPERATION

! ATTENTION !

-BEFORE SEVICING PUMPS-

1. Shut down or disengage the pump power source.
2. Shut down all pump accessory equipment.
3. Relieve or "bleed off" all pressure from the pump fluid cylinder.

Failure to shut down power and relieve pressure from the pump before servicing can result in serious personal injury and property damage.

WARNING

EXPOSURE TO THIS EQUIPMENT DURING OPERATION MAY CAUSE DEAFNESS. SUITABLE EAR PROTECTION MUST BE WORN.

WARNING

Failure to observe the **WARNINGS** can result in **property damage, serious bodily injury or even death!**

STANDARD OPERATING PROCEDURE
MUD PUMP OPERATION

PPE REQUIREMENT: Ear protection, Safety glass, safety shoes, work gloves, Hard hat.	

Purpose: Mud pump operation for fluid circulation.

Initial condition: SCR to mud pump motor assigned, Mud pump control at driller's Control mode, all auxiliary motors on auto mode, charger pump On auto mode, mud pump suction lined up with charger pump, discharge and suction dampener charged as per norms, mud pump by pass line closed, mud pump discharge lined up to rig floor, sufficient power generation as per load requirement available.

Reference documents: 12-P-160mud pump User Manual

Job hazards: As per the rig hazard analysis manual.

Tools required: None.

Operations:

1. Mud circulation through drill string or annulus.

Steps	Key points
1. The following checks to be carried out:	1. Bring the pump to speed gradually to

- Rig air pressure – OK.
- Liner cooling water level - sufficient quantity.
- Check liner and piston clamps –ok
- Suction pr. on the charger pump – ok
- Line up from mud pit to mud pump - ok

2. Line up mud pump discharge to drill string or annulus.
3. Assign SCR to mud pump motors.
4. Check the indicator for the following are illuminated:
   1) Power packs: Sufficient nos. (red)
   2) SCR : Corresponding (green)
   3) SCR Assigned : corresponding (yellow).
   4) Mud pump assigned: Corresponding (yellow).
   5) M/P field assigned (Yellow).
   6) M/P Aux assigned (Green)
   7) Charger pump running (green)
5. Rotate fine throttle clock wise.
6. Verify :
   *Both motors – Input current ok.

allow velocity of the fluid in the suction to match pump requirements.

2. When operating at low SPM between 5 to 30SPM, open cross head hand hole cover to check the oil supply.
3. When starting against pressure prime the pump by operating through by pass line in order to avoid air lock.
4. Suction dampener air pressure – 15psi
5. Discharge dampener nitrogen pressure – 400 to 650psi.

7. Rotate fine throttle clock wise and bring pump to required SPM.
8. Verify:
   - Pressure as anticipated.
   - No pounding sound on the pump to indicate cavitation.
   - Popping sounds audible on all the valves on the mud pump manifold.
9. To stop pump, slowly rotate fine throttle anti-clockwise.

Warning:

1. The relief valve outlet Must Not be connected to the pump suction line. Equipment damage or injury can happen.
2. **Use of high pressure air source or nitrogen bottle to charge the suction dampener can result in equipment damage and serious injury to personnel.**
3. **Relief valve setting should be – Piston pressure rating + 10%.**

www.ingramcontent.com/pod-product-compliance
Lightning Source LLC
Chambersburg PA
CBHW021007180526
45163CB00005B/1915